In 1900-1 Napier built a few vans on their 12 horsepower two-cylinder car chassis. They were reminiscent of the Daimlers built from 1897 and must have been quite successful, judging by one that was still running in 1905 after covering 70,000 miles (113,000 km) and having cost only £10 in repairs.

OLD DELIVERY VANS

Nick Baldwin

Shire Publications Ltd

CONTENTS

Published by Shire Publications Ltd, Midland House, West Way, Botley, Oxford OX2 0PH.
Copyright © 1987 by Nick Baldwin.
First published 1987; reprinted 1992 and 2009.
Transferred to digital print on demand 2011.
Shire Library 187. ISBN 978 0 85263 845 3.

Printed and bound in Great Britain.

Editorial consultant: Michael E. Ware, Curator of the National Motor Museum, Beaulieu.

All photographs and other illustrations are from the collection of the author.

British Library Cataloguing in Publication Data: Baldwin, Nick. Old delivery vans. — (Shire album; 187). I. Vans — History. I. Title. 629.2'23. TL230.
ISBN-10: 0-85263-845-0.
ISBN-13: 978 0 85263 845 3.

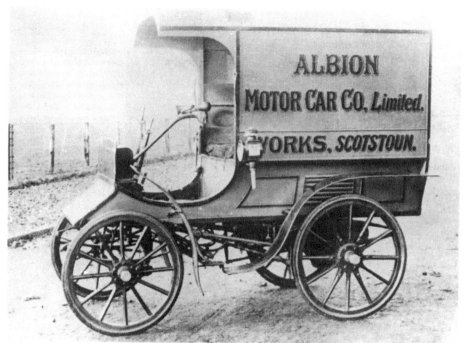

Albion's first van of 1902 had a horizontal twin-cylinder 8 horsepower engine and steering by tiller. It could carry half a ton. Albion was so far from the lucrative London market that it entrusted sales there to the Long Acre Motor Company, a firm which later made the rival Lacre van.

THE EARLY YEARS

Motorised delivery vans appeared early in motoring history, in the late Victorian age, and in Britain the first was built by Daimler of Coventry in 1897. However, like the cars on which most of the early ones were based, they became popular only after the first generation had been tried and tested. Until 1905 the few users, or potential users, of commercial motors had only the industrial sections in car magazines to consult about the latest developments. Then the publishers of the *Motor* brought out *Commercial Motor,* which was followed by *Motor Traction* (later *Motor Transport*) from the publishers of the *Autocar.*

Vans had long existed as horse-drawn vehicles and in many instances steam traction engines had taken over the hauling role, particularly in the household removals trade. From these traction engines were developed steam wagons which typically could carry 3 to 5 tons.

Many were built as vans: others carried demountable containers on their platforms. Internal combustion engined lorries did not begin to challenge steamers in these higher payload ranges until about 1905, when Leyland began to offer petrol vehicles alongside its steamers. For loads of up to about 15 cwt (762 kg) the typical early van was mounted on a car chassis propelled by either petrol or electricity. The 1 to 2 ton range would be purpose-built as commercial chassis and several well known car makers like Thornycroft, Albion, Dennis, Simms, Napier, Wolseley, Arrol-Johnston, Argyll, Cadillac, Renault, Alldays and Onions, Gladiator and Thames supplied them. Before 1910, however, there was a growing band of specialist commercial chassis makers such as Straker-Squire, Glover Brothers, Lotis, Ryknield and Dougill whose main efforts were aimed towards business users. Very few built their own bodies and

Alldays and Onions made commercial vehicles in Birmingham from 1906 to 1918. This early example of one of their vans was used for carrying samples and contained 28 drawers on each side. It could carry 12 cwt (610 kg) and had a two-cylinder 12 horsepower engine.

this work was frequently undertaken by the traditional coachbuilders that existed in almost every British town.

Horse-drawn delivery vehicles predominated for house-to-house local deliveries well into the 1920s (and into the 1950s for urban milk delivery) but the motor vehicle was making serious inroads. For local parcels deliveries carrier bicycles were sometimes replaced by motorcycle combinations with trade boxes. There were also a number of purpose-built three-wheelers like the Phoenix Trimo, Rex, Auto-Carrier, Wall Tri-Carrier, Omnium and Phanomobil. The two wheels could be either at the front or the rear and the driver often controlled the vehicle from a seat at the back.

A variety of businesses used delivery

This Adams of 1907 was an early example of a semi-forward control van. It had a single-cylinder underfloor engine and easy-change epicyclic transmission, hence the Adams slogan 'Pedals to push, that's all' (apart from steering).

4

vans in the era before the First World War and initially it was the larger manufacturers of nationally distributed products who adopted motor vans. The affluent brewers, for example could afford to try the new development, although they seldom needed the covered load space provided by vans. Flour millers on the other hand, involved in another lucrative business, needed covered carriage and, as they carried dense loads, often used steam vans.

Department stores found that they could do multiple deliveries by motor van far more quickly than by horse van. By early 1912 Harrods had 63 Albions and Whiteleys had 22, which between them had covered 1.1 million miles (1.8 million km). Laundries were early converts as were fresh food growers, who used vans to rush fresh produce to the markets. Rather surprising early users of motor vans were the various railway companies, who found them cheaper to run than horse vans for distribution from goods depots. Another important van user from an early stage was the General Post Office.

The use of second-hand car chassis for building vans spread their availability to smaller businesses and many enterprising shopkeepers soon had a van or one of the growing range of light parcel cars. In 1914 these included the Morgan with a load capacity of only 1 cwt (50.8 kg) plus driver. It cost £100, which was £3 10s more than an Auto-Carrier from the firm later famous for its AC sports cars. A Girling 6 horsepower three-wheeler cost £130 as did a four-wheel Enfield Parcelette, which carried 5 cwt (254 kg) loads. The friction-drive Forest 7 cwt (356 kg) van was priced at £175, the recently arrived and British-assembled Ford T cost roughly the same and a 12 cwt (810 kg) Overland, imported from America, was priced at £250. However, the Overland was one of the few vehicles to have a relatively full specification, including a windscreen, as standard. General equipment on most of these vans was very basic and few came with lights or any sort of weather equipment.

Four-wheel brakes and self-starters were not available until after the First World War and several vans still had such archaic features as chain drive, solid tyres and oil lamps (if they had any lighting) though acetylene was also spreading to those commercial vehicles that were regularly used after dark. Dennis had pioneered worm drive on vans in 1904

Few early vans survive and this 1907 single-cylinder Cadillac is a rare exception. It is shown at the finish of the Historic Commercial Vehicle Society London to Brighton run which takes place on the first Sunday in May each year. It was typical of the cheap and reliable American vehicles available in Britain from the earliest days.

ABOVE: *A 1910 Belsize run under contract for a publishing firm by Auto Van Maintenance in London. Similar chassis were used for taxi cabs. Note the resilient, non-skid rear wheel treads and the wing nuts that allow the front wheel rims to be removed for puncture repairs or replacement — very necessary where horseshoe nails abounded.*

LEFT: *This three-wheeler parcel car was made in 1912 by Girling Motors Limited of Woolwich. Different driving ratios were provided by friction discs on some models and by epicyclic transmission on others.*

and this quiet system gradually ousted chains. Engines were mostly four-cylinder types after 1912, though amongst the makers of better-quality heavier vans both Thornycroft and Pagefield offered two-cylinder types and three cylinders were also popular. Twin- and even single-cylinder engines, often with noisy air cooling, were widely used on the parcel cars. Gearboxes were mostly of the sliding-mesh 'crash' variety, except on electric vehicles. The exceptions were mostly limited to manufacturers who favoured preselective gearboxes, such as Commer, and epicyclic transmission, like Adams and several American-inspired or American-built vans including the Ford.

Between 1910 and 1914 many later familiar names began to advertise their purpose-built petrol van chassis. Amongst them were Yorkshire, who had previously concentrated solely on heavy steamers, Siddeley-Deasy with their 1½ ton Stoneleigh, Vulcan and Rothwell, who both offered chassis for payloads as low as 15 cwt (762 kg), Unic (a French make handled by taxi specialists Mann and Overton), Halley, Pilgrim (an early convert to forward control on light vans, as were Consolidated Pneumatic and R and G), Belsize and many traditional heavy lorry makers.

As motor vans became more widely used, so their owners began to realise their advertising potential. To begin with the mere ownership of a motor was enough to signify an up-to-the-minute business. As traffic increased this impact was diluted and striking liveries began to be adopted. Colours tended to be pastel but attractive pictures of the goods being carried were often painted on the sides of the van. Even if there were no illustrations there would be carefully laid out sign writing. This would be painted freehand by a highly skilled coach painter, who would often employ gold leaf for the owner's name.

Early vans were well made by the standards of the time but were unnecessarily heavy in relation to the payload capacity and, very slow. 20 mph (32km/h)

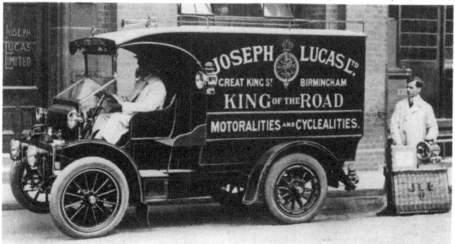

ABOVE: *From 1908 Austin made semi-forward control 15 horsepower four-cylinder chassis suitable for use as taxis or ¾ ton (762 kg) vans. The driving position allowed maximum body space within the overall length but was not widely adopted until the 1950s except on heavy commercials.*

BELOW. *Van drivers wore uniforms or heavy overcoats to keep out the chill. This chain-driven lorry is a Glasgow-made Halley with oil lamps and side curtains to keep out the draughts.*

ABOVE: *Battery-electric traction for stop-start local delivery purposes has always had a small following. This is a rare example from Tilling-Stevens, a firm which became better known for its petrol-electric hybrids, used mostly as buses in the 1920s.*

BELOW: *Dennis was the pioneer of worm drive on commercial vehicles in 1904, when it sold its first van to Harrods. The advantage of the axle was minimal power loss and much less wear and noise than on typical chain-driven vehicles. This is one of two vans supplied by Dennis to Battersea Dogs Home in 1909.*

ABOVE AND BELOW: *Even in 1913 most vans left their drivers exposed to the elements. This is a two-cylinder Straker-Squire laundry van and a chassis plan view of a four-cylinder version showing the typical features of two wheel brakes, right-hand accelerator pedal, separate gearbox with right-hand change and worm back axle, though other firms still preferred chain drive.*

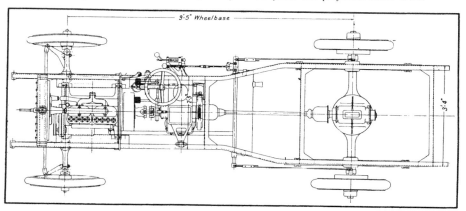

was the top speed for all but the lightest car-derived types and 12 mph (19 km/h) was the legal speed for the heavier types. Tyres did not last long, partly because of the poor roads and inevitable horseshoe nails (which explains why vans of over 2 tons had solid rubber tyres well into the 1920s), but otherwise vans were over-engineered and surprisingly reliable. A typical early Albion covered well over 100,000 miles (160,000 km) with nothing but routine maintenance. Future developments were indicated by the for-ward control types where the driver sat over the engine and at the front of the vehicle for maximum visibility and body space. One of the many manufacturers to adopt this feature was Glover Brothers, who in 1906 made a van with a body that could be drawn back, with the aid of a spiral screw, to reveal the engine. The tilt cab was ultimately to achieve the same objective. Renault, Arrol-Johnston, several American firms and Austin, who introduced a commercial chassis shortly before the First World War, favoured

9

Because of fuel shortages in the First World War attempts were made to run vans on town gas. Here a Humber and a Model T Ford (behind) carry their gas supplies in rooftop bags. Later vehicles towed or carried equipment to make the gas as they went along.

normal control but with the radiators behind the engine to reduce the risk of damage.

Another form of bodywork appeared in vehicles used as combined buses and delivery vans for rural areas and several enterprising shopkeepers had chassis that could be fitted with car bodies at weekends and van bodies during the week. Belsize offered a complete vehicle embodying these ideas, the Duplica, in 1914 for £260. Both these body ideas gained new adherents in the 1920s.

There were numerous direct American imports to Britain until May 1926 when 33¹/₃ per cent import duty was imposed. These Reos were photographed in 1924 in the livery of a well known operator. The Reo firm had been founded by R. E. Olds, previously responsible for the Oldsmobile.

Immediately after the First World War the Model T Ford outnumbered all other vehicles on the roads of Britain. The bodywork of this one has just been completed by Bayleys Limited. The American Model T had been assembled in Manchester since 1911.

MASS PRODUCTION

The First World War had a profound effect on road transport. Tens of thousands of returning servicemen had been taught to drive during the conflict. Many of them set up in business and took advantage of their transport knowledge. On high streets throughout Britain carrier bicycles, hand barrows and horses and carts were replaced by vans. These were often made from pre-war chassis or from military vehicles such as RFC Crossley tenders. In the range of heavier vehicles, thousands of former War Department 3 ton lorries flooded the market and gave a cheap start to many new haulage firms.

Amongst new vehicles the Model T Ford, which had been assembled for commercial purposes at Trafford Park, Manchester, since 1911, was the only one cheap enough to compete with the second-hand vehicles. It had the advantage of an epicyclic gearbox, which made

it extremely easy to drive for newcomers who were worried about the otherwise universal sliding-mesh crash gearboxes. These required considerable skill as it was necessary to double-declutch and match engine revolutions to road speed before another gear ratio could be selected. The Model T outnumbered all other vehicles on Britain's roads soon after the war but was soon challenged by a home-produced rival. This was the Bullnose Morris, which had started shortly before the war and was built on American mass production lines by a garage proprietor in Oxford. Its chassis was used as the basis for both cars and vans. Its price quickly came down as sales increased and by 1927 an 8 cwt (406.4 kg) chassis cost £122 10s whereas the equivalent Ford cost £105, having at one time been as low as £85.

William Morris then revolutionised the heavier vehicle market when he intro-

11

The Bullnose Morris was the first British-designed car to combat American imports successfully. Vans with the same engines were offered from 1914. From 1924 to 1931 the 'squashed bullnose' radiator of the vans differed from that of the cars. This is a van of 8 cwt (406 kg) capacity.

duced the 1 ton capacity Morris-Commercial in 1924 and soon several sizes of Morris van were competing against the British-built Ford and the dozens of different American and European makes on offer in Britain. The choice was bewildering and extremely damaging to large parts of the British motor industry. Eventually the government decided to protect the British producers and in May 1926 duty of 33⅓ per cent was imposed on imported vehicles. This ensured that Ford's American competitors, who already had strong commercial interests in Britain, began serious assembly and manufacturing operations there. This, in turn, enabled Bedford and Dodge to join Ford in the 1930s as British makes.

Meanwhile the success of imports and of car makers like Austin, Morris, Trojan and Singer, who had begun to offer vans, was having a disastrous effect on the traditional makers of lorries. These firms tended, each year, to make between a few hundred and a few thousand chassis, which inevitably were hand-built, labour intensive and expensive. They had typi-

cally offered van chassis from 1 ton or 1½ tons capacity upwards and Guy had even gone as low as 15 cwt (762 kg).

For under half the price of such a lorry it was possible to buy a perfectly satisfactory vehicle that was quite capable of the customary low mileage expected of most vans. For serious long distance work it made more economic sense to use a larger vehicle. Here the sales were relatively small and of little interest to the mass-producers, thus giving the traditional lorry makers an opportunity to recover after they had lost their sales of lighter vehicles.

Very few manufacturers built complete vehicles: they supplied the chassis and almost always local bodybuilders completed the bodywork. This meant that there was a wide choice of styles and liveries. Spray painting was unheard of and craftsmen painters and signwriters put all their skill into making distinctive vehicles. The publicity value of a striking van was accepted and tradesmen who were keen to be noticed sometimes had vans built in the shape of their wares or

A 1925 advertisement for one of the smallest vans made, with a capacity of only 2½ cwt (127 kg). It was of little more use than a sidecar outfit or two errand boys' bicycles, but it had far more novelty and publicity value.

A boon to the small trader

specialities. Dozens of different types were built, including facsimile toothpaste tubes, light bulbs, flower baskets, fruit, loaves, batteries, pens, bottles and vacuum cleaners.

In addition to the wide choice of bodywork there was a vast assortment of chassis to choose from, even after several of the importers had disappeared. The growing economic depression of the late 1920s ensured that every car maker began to look for additional sales. Many of those previously deterred by such an association with 'trade' were now glad of every sale they could procure. Even chassis with sporting pretensions, like Lea-Francis, Riley and Alvis, could be seen with van bodywork. All makes had to compete with the motorcycle and commercial sidecar combination that was quite popular for carrying the equipment of plumbers and chimney sweeps, for example. The horse van still monopolised local milk delivery work. Electricity had failed to fulfil its early promise, though

the number of 603 electric commercials in use in 1926 had almost trebled by the end of the decade. However, this was insignificant when compared with the 149,537 vans of 12 cwt (610 kg) to 1 ton capacity in use in 1928, the peak year of the decade. The under 12 cwt (610 kg) type was far less numerous (under 5000) until 1930.

Although the unconventional Trojan continued to sell reasonably well (primarily to one important customer, Brooke Bond Tea) even after Leyland had severed its financial links, the days of the unusual van were numbered from 1928, when the Model T Ford gave way to the Model A with a straightforward crash gearbox. The strange three-wheel parcel cars had been largely ousted by the Austin Seven and few had the features that had come to be taken for granted by van buyers as the 1920s progressed — weather protection, electric lighting and starting, shaft drive and car-type driving controls.

TOP: *This A model, from specialist lorry makers Thornycroft, was far more expensive than a similar mass-produced van.*

CENTRE: *A Morris-Commercial between two Unics in 1929. French Unics were quite popular and there were hundreds of Unic taxi cabs in London. Other European vans available included Renault and Fiat.*

LEFT: *The Trojan was one of the few bizarre van designs to find widespread public acceptance. On this 1927 van the solid tyres, the chain drive, the two-stroke engine under the driver's seat and the hand starting were already outmoded but Trojans changed little into the 1930s.*

ABOVE: *Large-capacity vans were used for long-distance trunking. This mid 1920s Maudslay is for 6 ton loads and has two wheel brakes and the solid tyres that were typical on heavy trucks until about 1930. Early in the 1920s solid tyres were used on vans of as little as 1½ tons capacity but as pneumatics quickly improved they were used higher up the weight range.*

BELOW: *In contrast, a few years later, this fast, low-built van, on pneumatic tyres, has the drop-frame Maudslay ML4 chassis normally used for passenger vehicles. It has four wheel brakes. Note the stylish car-type nickelled radiator and the roof rack for empty containers.*

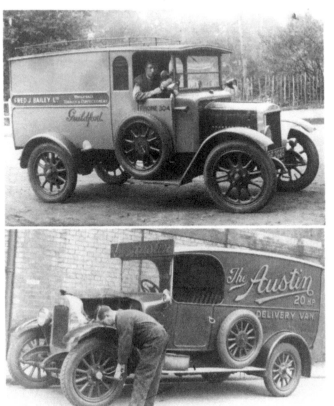

TOP: *Introduced in 1924, Morris-Commercial was the first cheap, mass-produced 1 ton van made in Britain. This 1928 light tonner cost only £175 in chassis form. It used the same 13.9 horsepower four-cylinder engine as many Bullnose Morris cars.*

CENTRE: *After 1918 Austin had intended to use 20 horsepower engines for all its different vehicle types. The Austin Seven on page 13 confirms that the plan had to be changed. This 20 horsepower van demonstrates the change to detachable wheels.*

BELOW: *A travelling shop mounted on a 1½ ton capacity Morris-Commercial chassis. The body and fittings were by W. H. Goddard of Oadby. The Morris-Commercial replaced the Model T Ford as the most commonly seen medium-weight van.*

ABOVE: *The Model T Ford gave way to the less unconventional Model A in 1928. For commercial purposes the 15 horsepower AF carried half a ton and the 25 horsepower AAF 1½ tons. This one, with an additional rear axle, could presumably carry at least 2 tons of books.*

There were several different types of product-shaped vehicles, though not many carried enough goods to qualify as vans. This 1929 Chevrolet did at least have rear doors on its two largest fountain pens. G. Wylder and Company of Kew Gardens were responsible for the ingenious coachwork.

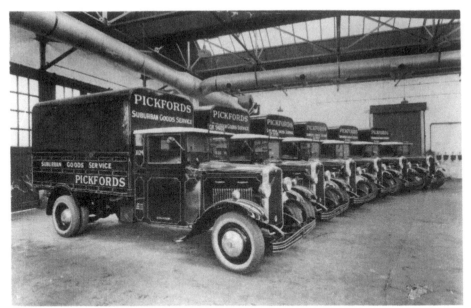

ABOVE: *Commer became an important force in the industry. In the 1920s it had been best known for heavy vehicles but the Rootes takeover had provided both money and car components with which to develop lighter models. From 1933 several of the lorries and vans from 1½ tons and upwards could have diesel engines.*

BELOW: *Heavy lorries continued to be specialised and expensive hand-built products, like this 1933· Albion. Both it and its drawbar trailer had flat beds but they are carrying demountable containers that could also travel by rail. By then cabs were fully enclosed and pneumatic tyres more or less universal.*

The most important newcomer to commercial vehicle manufacture in the 1930s, and to the van market in particular, was Bedford. It soon offered models for loads of 8 cwt (406 kg), 10/12 cwt (508/609 kg), 1½ tons and 2 tons. This is the largest type in 1933 complete with Luton head over the cab for light but bulky goods. The Luton head was so named because it had originally been useful for loads of hats from Luton.

THE 1930s

The decade opened with the worst economic depression of the century. Not only did this affect van users but it also thinned the ranks of commercial vehicle manufacturers. Of the many car makers who had diversified, few flourished. Bean and Singer both attempted to emulate Morris-Commercial but were defeated by the Depression. However, there were two major success stories that stemmed from this difficult time. One was that of Commer, which had come near to collapse but had been revived by the Rootes brothers in the late 1920s. Rootes also owned Humber and Hillman and through making components interchangeable between the makes they were able to make economies throughout their range of cars and light commercials. Their light vans might be pure Hillman or the heavier ones partly Humber, but all went under the name Commer until another lorry firm, Karrier, joined the group in the mid 1930s, when the same sort of component-swapping exercise again took place. Karrier specialised particularly in railway company, municipal and local delivery vans, for all of which oversize clutches and low-loading height were much in demand.

The Rootes Group was particularly enterprising with diesels and, as soon as the Perkins engines were available in 1933, Commers were offered with them, right down to the 1½ ton payload level. Admittedly there were very few buyers in the low weight range; total registrations of under 2½ ton capacity diesel vehicles in 1933 were nine, followed by 715 in 1934 and 703 in 1935. However, it showed the shape of things to come, especially when Leyland made their own diesels for all the vehicles in their range, including the little Cub, by 1934. The

Cub, new in 1931 as a 2 tonner, was a distant relative of the Trojan in that it was made in the Kingston-upon-Thames factory which produced the Trojan before production was transferred to Croydon in the late 1920s. The real significance of the diesel lay in its adoption by vehicles of 5 tons and upwards. Not only did fuel consumption improve but duty payable was much lower and reliability increased. Unfortunately diesel engines of the time had to be heavy and relatively low-revving, which did not suit the typical light van. Coventry Victor, who also made three-wheel petrol-engined vans, tried to remedy this from 1935 with a flat-twin diesel that could be substituted for the usual unit in the Jowett four-wheeler. However, few were sold.

The other great success story was that of Bedford. General Motors had done quite well with its Chevrolet light commercials assembled in Hendon and then Luton. These had included an increasing number of British components since import duties had been imposed in 1926. The acquisition of the Vauxhall car firm had given GM a major British manufacturing base and the two factors were brought together in 1931 to create the British-made Bedford. This started as a six-cylinder 2 tonner, but soon came to include 10/12 cwt (508/610 kg) and 1½ ton models again both equipped with powerful six-cylinder engines.

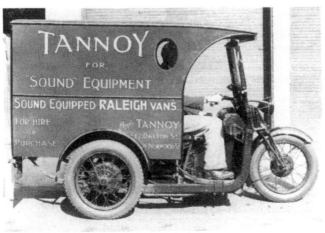

TOP: *Three-wheelers were a popular compromise between a motorcycle combination and a proper van and could be driven by a youth with a motorcycle licence for lower road tax than a four-wheeler. This is the 1931 Raleigh, redesigned as the familiar Reliant in 1935.*

BOTTOM: *The Austin Seven chassis continued as the basis for many vans in the 1930s. This one was used as a publicity gimmick by the makers of England's Glory matches and was fitted with small wheels and low wings. The driver was 3 feet 10 inches (1.17 metres) tall.*

ABOVE: *A wide selection of vans in the fleet of a confectioner and baker in the garden city of Letchworth in the late 1930s. Behind the little quarter-ton capacity 8 horsepower Fordsons are several electric local delivery vans and some unidentified petrol-engined types.*

BELOW: *Like its contemporaries, Morris-Commercial dispensed with aluminium for its radiator surround and adopted the American idea of a chromed ribbon radiator with the C-type from 1933. It was available with four- or six-cylinder engines and forward or normal control. 1½ ton to 5 ton versions were offered, and the T and the LT adopted its style and features.*

The 1935 Coventry Victor. 60 miles per gallon (21 km/litre) was guaranteed from its flat-twin 950 cc engine and stability was mysteriously described as being superior to most four-wheel light vans! The tax payable was only £4 a year.

The days when the typical cheap van was a rough and ready workhorse were fast disappearing and the average Bedford could outperform most cars and yet started at a mere £135 chassis price. GM also used Vauxhall to introduce their easy-change synchromesh gearboxes to the British market, and it was not long before this feature began to be included in light vans. Just as the Model T Ford's epicyclic system increased the appeal of the van for new drivers, so did the equally foolproof synchromesh.

Three-wheelers continued to enjoy a limited appeal because of their lower road tax rate. Raleigh, the bicycle firm, had become one of the best known names in this field, but sales were sluggish and they were glad to sell the complete department to T. L. Williams in 1935. As many of the components he took over featured an R for Raleigh, he decided to utilise them on a vehicle also beginning with R and the Reliant was born. To begin with, engines of motorcycle type were used but in 1938 Austin Seven engines with synchromesh gearboxes were adopted. The Austin Seven car and

Dodge became established in Britain in the 1920s and in time introduced completely British-made models. This is its 1937 van for 2 ton loads, from a range of ¾ ton to 4 ton models, all with six-cylinder engines.

The 2 ton Karrier Bantam was a very boxy little van. It was a close relative of Commer in the Rootes Group and normally concentrated on municipal vehicles and mechanical horses.

van were by then on the way out and in 1939 Reliant gained permission to make the Austin engine for its own purposes. It remained in production with various modifications until 1962.

A three-wheeler with the single wheel at the back was offered by Fleet in Birmingham, but, unlike on the Coventry Victor and others, the driver was placed at the back too! Perhaps the strangest of all the three-wheeler vans was the County Commercial Cars version of the Ford 8 with a tiny castor-like front wheel under the usual Ford radiator. Other conversions of cheap vans included those by Baico, who specialised in lengthening them and in altering Ford 8s, Austin 10s and the like to forward control.

Publicity vans could no longer be afforded by many van users but a few eye-catching types were built on outdated but mechanically sound vintage car chassis. Both Rolls-Royces and Bentleys came in for this treatment. A 6½ litre Bentley was found to be capable of carrying 1½ tons at 80 mph (129 km/h), but fuel consumption was presumably

heavy. Some stylish liveries and painstaking sign writing were still appearing on vans, though the process had been simplified and cheapened by the use of sprayed synthetics. The dozens of hand flatted and varnished coats of the typical coachbuilder of the 1920s had given way to quick-drying sprays that allowed bodies to be completed in a couple of days. This suited the mass-produced vans, many of which were built and bodied under one roof. Even more fundamental was the disappearance of chassis on some light monocoque cars and the arrival of independent front suspension shortly before the Second World War. Henceforth the van would have to develop separately, though wherever possible common engines, gearboxes and axles continued to be used to take full advantage of the economies of mass production. One of the first significantly different light vans was the 1936 10 cwt (508 kg) Morris, which had a very short bonnet made possible by moving the engine slightly to one side and partly into the cab with the driver's feet alongside it.

ABOVE: *Until 1952 the Ford light vans were known as Fordsons and consisted of the quarter-ton model shown here and a half-ton type, both of which stemmed from pre-war designs. Afterwards the name of the vans was changed to Thames, to fall into line with the larger Fords.*

BELOW: *The first of the new breed of forward-control light vans was the Morris-Commercial J type for half-ton loads, introduced in October 1949. It started with side valves but received an ohv engine in 1957 as the JB and was very popular with the General Post Office. This is the 1951 Commercial Motor Show exhibit with left-hand drive.*

Various high-powered chassis have been used as the basis of newspaper vans. This one is a 1947 Humber Super Snipe with 4.1 litre six-cylinder engine, but Ford V-8s had also been used and Wolverhampton-built Stars had been popular in vintage times.

1945 TO 1960

After the war ended in 1945 the priority for the motor industry was to switch to exports that could earn foreign currency. There was a pent-up demand for vehicles in Britain but shortage of materials and the need for exports meant that few were available on the home market. Former War Department utilities made in vast numbers by Austin, Standard and others were pressed into service as local delivery vehicles beside the pre-war vehicles that were still fit for service. The Land-Rover of 1948 was an example of the industry's response in providing utility vehicles with export potential, using Birmabright instead of scarce steel for its bodywork. The bodybuilding and specialist car making Jensen firm had tried a very light, integrally constructed aluminium van using Ford running gear before the war. Because it weighed under 2½ tons it was permitted to travel at 30 mph (48 km/h) instead of at the 20 mph (32km/h) of heavier lorries and after the war it was redesigned with a Perkins diesel as an extremely capacious 6 tonner. The traditional coachbuilders made wide use of aluminium and in the 1950s the new material, glass-fibre, also became popular, being light, rot-proof and suitable for short production runs.

New mass-produced small vans changed little until about 1950. By then even more light cars were of monocoque construction, so vans were able to develop independently along the most practical lines. Above 1 ton capacity they were mostly of forward control layout, though Bedford still favoured a bonnet. Forward control spread down to the 10 cwt (508 kg) Morris-Commercial J introduced in late 1949 and in 1952 Bedford adopted a semi-forward driver's position with its immensely successful CA. This was initially a 10/12 cwt (508/610 kg) vehicle though a 15 cwt (762 kg) version was added in 1959. It had a separate chassis but the body was of complete monocoque construction and featured sliding side doors. A 1.5 litre overhead

25

valve engine drove through a three-speed part synchromesh gearbox with column change to a hypoid rear axle. There were semi-elliptic rear springs and coils for the independent front suspension. In its seventeen year run, over 350,000 CAs were sold and they influenced the successive Austin/Morris 151/J2, followed by the small forward control Thames of 1957, the Standard Atlas of 1959 and the Commer 1500 of 1960.

Imports were rare during this period but one model that sold in appreciable numbers from 1955 was the Volkswagen 15/16 cwt (762/813 kg). It was strikingly different from established British practice in having a rear-mounted flat-four air-cooled engine under the loadspace. It earned a good reputation for reliability in service, though probably no more reliable than the average British water-cooled mass-produced van.

Electric vans continued to sell in very small numbers and motorcycle-derived types all died out with the exception of

the increasingly refined Reliant, which from 1955 lost its motorcycle forks and gained fibreglass bodywork. Other unusual types, like the two-cylinder Jowett Bradford, did not last for long, though the Trojan soldiered on until 1964. From about 1953 the Perkins P3 was its most usual engine and this brought diesel economy to the 15 cwt (762 kg) market. The mass-producers did not generally offer diesels until well into the 1960s, though at the start of the decade Commer offered a Perkins 1.6 litre four-cylinder diesel in its 15 cwt (762 kg) vans. At that time some of the larger Commers used 2.2 litre diesels made by Standard that were also to be found in Massey-Ferguson tractors. This same engine had also been used in a few commercial versions of the Standard Vanguard in the early 1950s.

There was far less van variety on the roads of the 1950s than at any previous time. Not only were there fewer manufacturers but most bodies were now

The Morris name was retained for vehicles derived from light cars as opposed to the heavier Morris-Commercials, though in 1951-2 the old Cowley name sufficed for this model. It is shown attractively liveried as a catering van and is based on the 1950 pattern Cowley car with torsion-bar independent front suspension and other refinements. Better known was the Minor-based van produced from 1953.

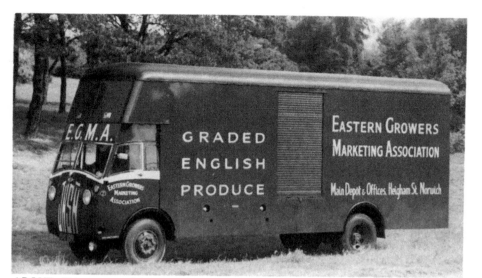

ABOVE: *This was one of the most advanced post-war lorries with a van body. It is a 1947/8 JNSN made by Jensen which had a very low unladen weight due to its aluminium semi-monocoque construction. It could carry 6 tons and had a Perkins diesel engine and Moss five-speed gearbox.*

BELOW: *Undoubtedly the most significant van of the 1950s and 1960s was the Bedford CA. New in 1952, it had a 1.5 litre overhead-valve petrol engine, three-speed synchromesh gearbox, semi-forward control and independent front suspension. A three-quarter ton version was offered from 1959 but before that the payload was limited to 10/12 cwt (508/609 kg).*

ABOVE: *The forerunner of the Bond light car was the Sharps Minitruck made in Preston, Lancashire, in 1953-5. This is an early model with a Villiers 197 cc single-cylinder air-cooled engine. Minimal vans like this and others such as the Jarc/Astra, Gill and Pashley were soon to disappear.*

BELOW: *Jowett made vans from the 1920s until they ceased production of all vehicles in 1953. All their vans had flat-twin water-cooled engines. Their post-war half-ton van was called the Bradford and, apart from the three-quarter ton Trojan, was one of the last small-production British vans to enjoy commercial success.*

standardised and often finished in one of a small choice of factory colours. Some were still sprayed to customers' individual requirements but even then the increasing cost of skilled labour was causing intricate signwriting to die out.

An obvious solution to the persistent design problem of leaving the maximum space available for payload was to keep all the mechanism at the front and there had been various attempts to interest customers in front-wheel drive vans. In 1954 the Turner gearbox and engine firm had offered a front-wheel drive light commercial and before that numerous continental firms had favoured the system though none had enjoyed any sales success in Britain. At the lowest level the little Citroen 2CV had been available as a van in France from 1951 but had reached Britain only in left-hand drive form.

However, this reluctance towards front-wheel drive was to change with the revolutionary Mini passenger car of 1959. The traditional rivals Austin and Morris had finally come together as the British Motor Corporation in 1952 and had slowly rationalised their model ranges. Alec Issigonis's independently sprung, front-wheel drive Mini with transverse engine was available with either badge. Placing the gearbox and differential in the engine sump gave an amazingly compact unit ideally suited for commercial use and from 1960 the Minivan began to influence the market for quarter-tonners. It also brought widespread acceptance of front-wheel drive by the British public, though, strangely, the average buyer of larger vans still preferred a thoroughly conventional rear-wheel drive layout. This held true into the 1980s and the new Transits of 1986 still eschewed front-wheel drive, though many of its continental rivals had adopted the feature.

The original Transit of 1960 set the standards by which all other vans of the era were judged, in the same way the Model T and Bedford CA had done. It was conceived at a meeting held between officials of Ford of Germany and Britain on 9th January 1960. The Germans

The Austin KB or Three-Way van, as it was better known, provided easy access from side and rear and was for loads of 1¼ tons. It used the same four-cylinder engine as the A70 car and the Countryman and provided maximum loadspace within a compact and manoeuvrable 7 feet 9 inch (2.35 metre) wheelbase. This one has bodywork and painting by Cunard Coachworks.

The Austin J2 of 1957 was a fully forward-control compact van. This model, known as the 152 Omnivan, had a 42 brake horsepower petrol engine and a capacity of 16¼ cwt (825 kg). The Sherpa was based on this model and an even closer derivative is produced in Spain.

wanted sliding side doors like the Volkswagen but in other respects their ideas coincided with the British and led to a far larger van than had hitherto been built for 12 cwt (610 kg) loads and allowed the same shape to suffice for models of 1½ tons capacity and over. The success of the Transit can be seen in the two million sold in its 21 years existence and by the numerous other vans that have adopted its concept.

Based on the Hillman Minx of 1956 onwards, the Commer Cob with 1265 cc 35.5 brake horsepower engine was for 7 cwt (355 kg) loads. Commer's 8 cwt (406 kg) Express Delivery Van had the larger 1390 cc engine of the De Luxe Minx and Californian.

FURTHER READING

Several lorry books touch on the subject of vans and the following is a list of some useful titles that may still be obtainable new or second-hand from specialist bookshops and transport flea markets.

Aldridge, J. & Thomas, A., Licensed to Carry. Motor Transport, 1976 (the story of the Letland Group.)
Baldwin, Nick, Heavy Good Vehicles 1919-39. Almark, 1976.
– Kaleidoscope of Lorries and Vans. MHB (Haynes), 1979.
– Observer's Book of Commercial Vehicles. Frederick Warne, various editions.
– Observer's Book of Trucks. Penguin, 1986.
– Pictorial History of BRS. MHB (Haynes), 1982.
– Vintage Lorry Albums/Annuals. MHB (Haynes), 1979 onwards.
Baldwin, Nick & Ingram, Arthur. Light Vans and Trucks 1919-39. Almark, 1977.
Edwards, Harry. Morris-Commercials –The First Years. Morris Register, 1983.
Georgano, G.N., Complete Encyclopedia of Commercial Vehicles. Krause, 1979.
–World Truck Handbook. Jane's, 1983 and 1986.
Marshall, Prince. Lorries, Trucks and Vans 1897-1927. Blandford, 1972.
Miller, Denis, Illustrated Encyclopedia of Trucks and Buses. Quantrum, 2002.
Whitehead, R.A., Kaleidoscope of Steam Wagons. MHB (Haynes), 1979.

There are several commercial vehicle magazines, some of which cover historic machines. Nick Baldwin writes mostly for *Classic and Vintage Commercials* published by Kelsey, Cudham, Kent, TN16 3AG which also produces *Classic Van and Pickup*.

RALLIES AND CLUBS

Vans have always had intensive use and not many lasted for more than ten years. As a result few originals survive although in recent years several old car or light commercial chassis have been restored as vans.

At the old vehicle rallies that take place on most weekends outside winter there will be several vans and the Historic Commercial Vehicle Society runs, including the famous London to Brighton on the first Sunday in May, the Trans-Pennine in early August and the Bournemouth to Bath on the first Sunday in September, always have a good selection.

Numerous one-make car clubs have sections catering for van owners, such as the Morris Register (contact Harry Edwards, Wellwood Farm, Lower Stock Road, West Hanningfield, Essex). Most van owners belong to the Historic Commercial Vehicle Society (details from Iden Grange, Cranbrook Road, Staplehurst, Kent TN12 0ET) or the Commercial Vehicle Road Transport Club (8 Tachbrook Road, Uxbridge, Middlesex UB8 2QS). Both of these clubs have regular newsletters containing information about events, availability of spares, changes in laws relating to old vehicles and histories of vehicle manufacturers and operators.

PLACES TO VISIT

These museums are known to have vans and related vehicles on exhibition. Intending visitors are advised to find out times of opening before making a special journey.

British Commercial Vehicle Museum, King Street, Leyland, Preston, Lancashire PR5 1LE.
    Telephone: 01772 451011. Website: www.bcvm.co.uk
Cotswolds Motor Museum, The Old Mill, Bourton-on-the-Water, Cheltenham,
    Gloucestershire. Telephone: 01451 821255.
    Website: www.cotswold-motor-museum.co.uk
Coventry Transport Museum, Hales Street, Coventry, Warwickshire CV1 1JD.
    Telephone: 02476 234 270. Website: www.transport-museum.com
East Anglia Transport Museum, Chapel Road, Carlton Colville, Lowestoft, Suffolk NR33
Grampian Transport Museum, Alford, Aberdeenshire, Scotland AB33 8AD.
    Telephone: 01975 562292. www.gtm.org.uk
Haynes International Motor Museum, Sparkford, Somerset BA22 7LH.
    Telephone: 01963 440804. Website: www.haynesmotormuseum.com
Heritage Motor Centre, Banbury Road, Gaydon, Warwickshire CV35 0BJ.
    Telephone: 01926 641188. Website: www.heritage-motor-centre.co.uk
National Motor Museum, John Montagu Building, Beaulieu, Brockenhurst, Hampshire
    SO42 7ZN. Telephone: 01590 614603. Website: www.beaulieu.co.uk
Science Museum, Exhibition Road, South Kensington, London SW7 2DD.
    Telephone: 0207 9424000. Website: www.sciencemuseum.org.uk
Science Museum outpost, Wroughton, Swindon, Wilts SN4 9NS.
    Telephone: 01793 846200.

Printed and bound by CPI Group (UK) Ltd, Croydon, CR0 4YY

11/10/2024

01043563-0001